SOMETHING INCREDIBLY WONDERFUL IS IN THE WORKS: SELECTED LETTERS OF LESLIE LACKMAN 2007-2010

EDITED WITH AN INTRODUCTION
BY
ALLEN GREEN

A Blue Logic Publication
www.bluelogic.us

ISBN 978-0-6922639-5-2
Printed in the United States of America

I get this feeling that something really, really, really, REALLY genuinely WONDERFUL and surprising for everyone…is on its way, maybe already here even, just taking its time to solidify into something that people bumming around in bodies can recognize.

Contents

Introduction

I Dream of Leslie

Shortly after I decided to accept the challenge of publishing these letters of Leslie Lackman, she came to me in a dream during the night of November 30, 2012. What I most remember is how the two of us were looking into each other's eyes. I can't remember a dream where another's face and eyes were so clear, where we were so connected. Leslie was talking to me about something, although I don't remember what. But she was teaching me something. Even in the dream, it did not feel like anything specific, yet I knew it was significant. I remember listening intently, absorbing this private, one-on-one teaching, all the while maintaining constant eye contact. I remember thinking something like: "She's gone from the physical world, but I can still learn a lot from her. It will never be too late for that. She will continue sharing and teaching."

1% of 1% of ...

I really knew only a small part of Leslie. I was mostly unaware, until after she died, of her interest and talent in writing poetry, in ceramics, in dancing. I knew she enjoyed singing, flower arranging, and carving imaginative and unique Halloween pumpkins. And I also knew that just about everything in the world interested her. I knew that she wanted to understand as much as possible about science and medicine and politics and history. But she was not at all satisfied with the learning the average person might get from newspapers, magazines, encyclopedias, or any other source readily accessible to the "masses." She needed to go deeper, not even to be content with being amongst the

"1%-ers," or even 1% of 1%. What she understood was that there is no end to how deep one can go, how many connections can be made, how high one can climb above the understanding of the average person.

I Think, Therefore I Am

So what can I say about Leslie, based on her writings, and on various discussions between us?

We can start with this:

> *"The first step in manifestation is: twinkle in the eye."*

This is quintessential Leslie: The world is as good a place (or as bad) as we imagine it to be. A very important subject for Leslie, and the one represented the most in the letters selected for this book, concerned what Leslie called the "informational realm," with the interconnected "Law of Attraction." I don't want to say much on this subject, because Leslie was way ahead of me in this and explained it so much better than I possibly can. Leslie believed with all her heart that our thoughts take root in the informational realm, and then create a corresponding expression in the physical world and in one's life. Anything good and positive that happens to one is the result of projecting a positive "blueprint" into the "informational realm."

Johann Goethe actually had a similar understanding and expressed it like this:

> *Until one is committed, there is hesitancy, the chance to draw back. Concerning all acts of initiative and creation, there is one elementary truth, the ignorance of which kills countless ideas and splendid plans: that the moment one definitely commits oneself, then*

Providence moves too. All sorts of things occur to help one that would never otherwise have occurred. A whole stream of events issues from the decision, raising in one's favor all manner of unforeseen incidents and meetings and material assistance, which no man could have dreamed would have come his way. Whatever you can do, or dream you can do, begin it. Boldness has genius, power, and magic in it. Begin it now.

The Magnificent Ten

Here are a few more thoughts about Leslie's character, the quality of her thinking, her outlook on life, and her relationship with the world:

• Leslie always found her own way of expressing something. She shunned canned phrases and, instead, wrote prose like a poet. She was always looking to play with words to find new twists and turns of expressing herself:

"...there are many languages that require not intellect but heart,...and intuition, and the being-ness of being connected to the great, the mysterious, the totally within and without, the fabric of all, the one and only, the simply profound and profoundly simple and ever present—One."

• What subjects was she not willing to tackle!? From astronomy to healing, politics to alternative energy, economics to sex, esotericism to aliens:

"Maybe there was a migration of Greys and others from Mars to Earth, just before the comet-that-became-Venus fly-by that stripped Mars of its atmosphere and water."

• Regardless of how negative a particular subject, or a certain realization about the state of the world, she always tried to look for a positive outcome. The future, on the level of ordinary life, would certainly be better than the present:

"In spite of everything we know, I still have the feeling that something incredibly wonderful is in the works."

• She had a wonderful ability to bring together ideas from very different eras and cultures, like connecting "Robert's 30 work 'I's" with Francis Bacon's Spear-Shakers.

• She was extremely well read.

• At any time, she was ready and willing to leave the Earth behind and embrace the highest celestial thoughts:

"To have a mind—whether one is conscious or not—means that we are renting a piece of God's mind."

• She lived at the leading edge of understanding life, reality, how it connects and how we connect to it all:

"It is becoming more and more clear to me that what we have been taught to call "reality" is actually only the readout of information patterns and energy patterns..."

• She could be brutally honest and unflinching in seeing the world as it is with all its power struggles and greed :

"'A New Birth of Freedom'...oh dear, they're using that hypnotic word again. We'll kill anyone for 'Freedom'..."

• She was so well versed in so many areas, and went so deep into understanding and connecting, that she could simplify and explain otherwise difficult ideas.

• She could be downright zany and crazy, right out of Wonderland, imaging a much better world, a world with unlimited free energy, no debt, and real healing solutions available to all:

> *"This morning, I woke up and went outside and I found a pile of money in a trunk that said 'Eternally valuable because you are.'"*

In The Beginning

The letters reproduced in this book were originally emails. For several years, Leslie corresponded with some 15 people in a sort of "email discussion group." Thus a number of her letters begin by addressing one of the group members (Charles, John, etc.), who had written the email immediately preceding Leslie's response. I chose not to include letters from anyone else for two reasons: First, there may actually have been a chain of letters leading up to Leslie's response, which would have been too lengthy to include here. Secondly, this book is about Leslie, her thinking process, her heart and mind, her humanity. I did not want to deviate from this aim by including thoughts from other people, or by focusing too much on a particular discussion point.

Categories

As far as the grouping of Leslie's letters, it became apparent that most of the letters fell into one of three categories, although these categories overlap in many cases. They are:

1. Healing-related subjects,
2. The invisible world, including the "information realm" and the Law of Attraction, and
3. Politics and world affairs, i.e. the affairs of men in power and the games they play.

Letters not clearly falling into one of these categories have been grouped under the heading *Miscellaneous.* Within each of these four categories, the letters have been organized chronologically.

If It Ain't Broke, Don't Fix It

I have tried to minimize any editing of Leslie's letters. Leslie had her own style of writing, generally very personal, highly emotional, and unique in how she expressed herself, even making up words when required. I left intact everything that I could. Maybe once, she omitted a preposition, which I inserted. Maybe 10 times, I added a comma here and there to make the meaning of a sentence a little more clear. In all these letters, she misspelled only a couple of words, which have been corrected. And that's all. The other 99.9% of these letters is exactly as Leslie wanted them to be.

Final Instructions

The aim of publishing these letters is not to promote a point of view, or to force exposure to any particular subject. This book of letters should definitely not be seen as subtle propaganda. However, should the reader be attracted to a particular subject, and even desire to research something in more depth, I'm sure Leslie would not only be pleased, but would feel some reward for her own uncountable hours of reading, research, thinking and sharing.

But if the reader is not particularly attracted to any of the ideas expressed in these letters, there is still a considerable amount to be gained by observing a beautiful mind at work and at play. For Leslie was at once both profound and playful; thoughtful and unrestrained; poetic and raw. Her writing followed no conventions. She made up words and expressions as befitted the idea and emotion she needed to express. No subject was too low or too high. There was always more understanding to be garnered, and additional inner connections to be uncovered. Leslie was all about being open to anything new, and finding how it connected to everything she already understood. She shunned nothing, even the most coarse realizations of how the world worked. But she never let such impressions stay with her at their low vibrations: Whenever possible, she strove to transform these base, worldly realities into something more beautiful, more hopeful. This was one of her great inner skills.

To repeat: The publishing of these writings is not for the purpose of you, the reader, learning anything in particular. Don't read them in that vein. The purpose of this little book is to give you perhaps a view of Leslie that you never had. It's to show her wonderful intellect, her humanity, her multi-faceted interests, which extended far beyond the bounds of her written words. For she was also an artist and a potter and a poet, and so much more. Yes, learn a little if you so choose. But please, please leave all judgment behind starting right now. Become light, like Leslie. Allow her to take you, as she has taken me, on a very special journey, a journey through all cultures, all sciences and religions, back in time as well as throughout the Solar System. If you allow her, she will show you astounding vistas and amazing beauty that you may never have seen before. So buckle up, relax, sit back and enjoy the ride.

"Have to keep looking inside ourselves for sanity and love."

Allen Green
Oregon House, CA
July 2014

The Letters

Politics and World Affairs

I don't believe that oil comes from squished dinosaurs.

1-5-07

Hello dear fringe friends,

War, and all its alleged justifications, of course including our so-called "war" on "terror," propagate themselves like mushrooms with words and phrases whose validity goes unexamined by the majority of people once the foundational event--911 in this case--has been perpetrated, lied about, reacted to, died for, marketed in headlines, movies, television shows, video games, toys, and books, and referred to directly or indirectly in newscasts daily and hourly. I'll just bet there is a graph somewhere, the Propaganda Effectiveness Cycle Chart, that shows how long to report bad news (a certain number of minutes or days) and when to interject some good news so that everyone's guard goes down allowing the next wave of bad news (more car bombs, more suicide bombers, more dead people) to go deep again and make it all "REAL."

Today at the flower studio, someone brought her 5-year-old in. He asked questions about the damaged ears on the Nefertiti bust, which was there at the time, I started explaining something about copying the original statue, and within a minute he was talking about how you could take a knife and stick it into someone's head and just scrape off a face and even ears if you wanted to. I said, "What are you talking about? That's really violent and bloody, what you are saying. Why are you saying that?" He said enthusiastically, "That's what they do in wars." Is it the media, or would little boys get those ideas even without the media?

L

§

2-24-07

Dear ones,

This line sets the stage for a whole new level of holocaust/ genocide (Somewhere someone said that "holocaust" means "sacrifice."):

"It's important to understand when you start to deflect an asteroid that **certain countries** are going to have [to] accept an increase in risk to their populations in order to take the risk to zero for everyone," said Dr Schweickart

So some big thing, publicized as an asteroid but which is some weapon destructive enough to seem like an asteroid after it hits, falls on some country full of people on the list for extermination/population reduction (the way hurricanes in Forida by strange coincidence put only the "white trash" mobile home neighborhoods through the blender). Oops, awfully sorry, well, you know how it goes with asteroids, folks. We can only *begin* to *deflect* them. After all, it WAS an asteroid, wasn't it. We can't control them, we can only begin to deflect them.

It reminds me of this, which sets the stage for totalitarianism:

Loss of freedom is the price to pay for greater security in this age of terrorism. So you don't have any constitutional rights anymore? Would you rather have a terrorist attack?

You mean like the terrorist attack that my government orchestrated in New York?

Well, actually, I'd rather have a government that is not a terrorist organization...

Leslie

§

Words, images, and symbols are powerful because they act in the informational realm, where the blueprints, the Original Ideas, the matrices of being are.

3-20-07

Dear friends,

So, essentially what this article (or at least the first part) says is that the chemtrails are good, they are being used to STOP global warming. Maybe with some health-harming things like barium and aluminum, but, although we would have liked to have a say in what they used to "increase the albedo," essentially, folks, the chemtrails are there in the sky to protect us in the long run.

I don't believe it. I think it's just a way to fool all the scientists. Get them all involved, make a big frigging deal about stopping global warming, then do the most amazing thing: IMPLEMENT global warming with chemtrails while making everyone believe that they are stopping it. It's brilliant.

The chemtrails are contributing to, if not outright CAUSING global warming. They are not mitigating it. But just like the words "freedom" and "democracy", the hypnotic words here, for those that will hear/read/write/be involved with these "reports," seem to be "mitigation" and "increase albedo".

I don't believe that oil comes from squished dinosaurs (because if it does, that means that the La Brea tar pits are lumpy potatoes) and I don't believe that covering the earth with a blanket (of haze) makes it cooler. It makes it WARMER. To prevent heat from accumulating (in a greenhouse, a car, a room, or on the earth), light must be blocked completely, because if it is not, it reflects back as heat, and cannot pass through the glass/walls/clouds the way the light does. That is what greenhouses and hot cars with the windows closed are all about. And the chemtrails, no matter how many of them there are, do not block the light completely.

Leslie

§

12-22-07

Me neither. Haven't had a vaccination for years. They're bidding on the bird flu...

Apropos of vaccines, I was speaking with a student recently who is taking courses related to a medical profession, and he said that it was mandatory for him to get a vaccination in each arm (hepatitis, I think), and both arms feel adversely affected--weaker--since getting them.

Generally speaking, for my own welfare, rather than letting my mind wander anymore into negativity about all that goes on in this world, I have decided to look upon the Illuminati, the New Worlders, the Bushes, the Rothschilds, et al., et al., et al., with respect, awe, and admiration for their unparalleled expertise in manipulating what people think and believe and for perfecting the machinery of perpetual, unlimited income, with nearly every single person on the planet putting money into their pockets, and I pray that their hearts may be changed and filled with the One True Light. Nice phrase, ay? I just mean something really really positive and wonderful like ...

Joy, peace, love, and light.

Leslie

§

3-18-08

I think anything that the sponsors of 911 want to do can be done just as neatly as 911 was because everyone is just "doing their job." There are mercenaries, assassins, demolition experts, you name it, that can be hired for whatever they need, and everything is done on a strictly need-to-know basis--heck, on 9/11, even though Bush was in on it, I don't think he *knew* it was happening that day, so when he said to the press "I'm angry" (not sounding angry at all), I had the feeling that he was angry that they did it *that particular day* without telling him, maybe hoping that his phony responses would then sound more authentic (but they didn't).

No, even with more people becoming aware and so on, the only ones who could stop it are the ones who are hired to do the job--THEY would have to be aware, somehow miraculously freeing their minds from the seeming realities of the whole crazy game. They are the ones who would have to not shoot, not kill, not bomb, not fly the planes, and not push the buttons. They would have to not follow orders, because they are the ones sitting at the controls of the weapons. They would have to stop the wars and "terrorist attacks" that they are being recruited, hired, trained, and paid to wage.

Wouldn't it be loverly if that happened? All the guys, sitting at all the controls, knowing that something was deeply wrong with the whole thing, decide to refrain. They turn off the "Burn, baby, burn" R&R soundtrack playing in their helmets. They don't plant the next car bomb (or they make it a dud). They ignore the threats from some rabid "commanding officer." They realize that murder and destruction are not freedom and democracy. And then, being the highly trained experts that they are, they permanently disable all controls that deploy the weapons. What a lovely thought.

Leslie

§

People accomplish remarkable things because they persistently believe they can, and act in support of their belief.

8-24-08

A woman president--dream come true.
A man of color for president--dream come true.

To me, it just seems like a way to "harvest energy." You know how it goes in the Middle East: "Today they talked about peace." Hopes rise, hearts open. One or two or three days later, another bomb goes off. People get killed. Energy harvested. Then more fighting, more same old same old. Then the same sequence is played again thousands of times, only with slight differences in timing. Heartbreak every time. Energy harvested every time. Human spirit (which may be the target, the most threatening thing to NWO) bludgeoned again and again.

So, presidential "elections." It starts looking like "dream come true" time. Hopes rise, hearts open. How will the energy be harvested? Some incident that dashes it all to bits, maybe. Another 9/11 and this time, martial law. And even if that does not happen, and one of the dreams-come-true gets "elected" (what a joke), he/she has already had their interview with the Bilderberg group, right? and will be told what their agenda really is, when it's time for her/him to know.

However, in the meantime, something far less sinister may be afoot, and may be gathering energy also.

Leslie

§

8-28-08

Don't forget about the tactic of the one-two punch, the first part of which has taken place at the Democratic convention. It is a tactic in trauma-based mind control: open their hearts and then break them. With children, they let a child get close to a pet, for example, then kill the pet.

So, Obama is our new pet: the democratic candidate... historic... first (barely) "black" candidate ever... tears of joy everywhere... that's #one: open their hearts, raise hopes.

Seems possible that Iran could go as scheduled, but the terrorist organization known as U.S. military forces will make it look like it's all Russia's fault---that's #two: break their hearts with ....more war? Maybe an assassination of Obama, a la JFK and Bobbie? Both? Then someone else steps in... and perhaps some way down the line, having set the precedent for a non-American-born candidate, Schwarzenegger will step up to the throne and officially announce the coming of the Fourth Reich in America. Hooray.

But then, in other places, in other minds, in other dimensions, something far less sinister is taking place...

L

§

It's amazing that "scientists" still publish such juvenile things about the impossibility of things that are a cinch for God.

9-25-08

The players play their parts. These guys are really good script writers, I must say. Surprising twists and turns, gripping suspense, raise your hopes, dash your hopes, make you feel the danger... wow, what a show. I'm going to tell my friends about it. "Allow me to issue and control the currency of a nation, and I care not who makes the laws." or something like that. Mr. Rothschild. Let's all give a warm welcome to the Final Financial Solution: the AMERO!

Well, that takes care of two out of the ten regions planned for in the NWO global plan of rulership, ay? Whew, at least it's not going to be a terrorist attack. It's a financial attack, cancelling the "election", and dubbed "worse than what happened in the 1930s," so therefore, we should all be very happy when they come up with the amazingly creative and out-of-the-box "solution": turning North America into one big happy country, controlled by Rothschild's law. Awww, poor Americans, why are you so surprised? What do you think just happened in Europe a few short years ago? Do you *still* think that no one is smart enough to pull something like this off? Do you still think that the word "conspiracy" should always be accompanied by the word "theory" and a sneer? Well, bumbling or not ... they just took our country away.

At the same time, for no logical reason, well, a slightly logical reason, I get this feeling that something really, really, really, REALLY genuinely WONDERFUL and surprising for everyone---including us smart cookies who have all the melodrama and phoniness figured out---is on its way, maybe already here even, just taking its time to solidify into something that people bumming around in bodies can recognize.

But if it gets bad for the bodies, let's hope that there's more to life than having one.

love, *Leslie*

§

10-24-08

Well, I have my canned fish, beans, and TP so far. Need more containers for water. And matches and candles... but I keep having this feeling that, even with this awful stuff happening, there is an equal or greater amount of good things happening and that, while many people who don't read what we read will be surprised at how bad it gets, we may be surprised at how good it gets, in some way...

love, *Leslie*

§

1-11-09

“A New Birth of Freedom” . . . oh dear, they’re using that hypnotic word again. We’ll kill anyone for “Freedom.” We’ll bomb the crap out of any country, and any school, school bus, hospital, church, or water supply system in any country to bring them “Freedom.” Because “Freedom” justifies everything. “Freedom” gives us carte blanche. “Freedom” is sacred. “Freedom” cannot be scrutinized. The inaugural address “A New Birth of Freedom” delivered by a black guy is guaranteed to deepen the trance.

Leslie

§

It’s important as individuals to not worry, and be happy, like all the children and animals that don’t listen to the news…because mind creates blueprints.

8-27-09

The powers-that-be know that *people are satisfied with having emotions* about it all, rather than knowing or understanding what’s really going on. People don’t *think* beyond their emotions, and the PTB provide those emotions on a rotating basis with their “news.” It’s part of trauma-based mind control. When people talk about what they experienced in relation to a traumatic or tragic event (Viet Nam, WWII, concentration camps, 9/11, etc.), that's the only "reality" they are interested in, because it produces in them (us) the most potent and intense energies. After that energy has imprinted us with 3rd and 4th state memories, there is almost no possibility of allowing the king of diamonds to turn up new evidence that would reveal the motives, agendas, tactics, and manipulations that have provided us with that "reality."

§

4-15-10

Next to sunlight, water, and empty space, sex energy is the most abundant thing on the planet. How can any healthy creature be completely celibate? And if they are celibate, how can they stay healthy emotionally, mentally, or physically? Sex energy doesn't just go away, but it can get distorted. So, with an entire religion expecting its leaders to be celibate . . . maybe it's an unhealthy organization to begin with. Time for change. Defrock them all, start a different kind of religion.

Leslie

§

4-16-10

Funny how business just goes on and on. No one notices that almost nothing of human value is ever gained in the exchange of bombs, missiles, bullets, etc.

Let's have another Good News Bulletin, like the one on April 1. *(See letter of 4-02-10 in the next section - Ed.)*

hugs, *Leslie*

§

5-12-10

I agree that we are being set up--by movies, books, news, think tanks--and Allen probably has many more observations of how it's happening than I have. You know, I think that even Disneyland was setting us up. I have not been there for decades, but from when I was a child, I remember the exhibit of international dolls singing "It's a small world after all, it's a small world after all, it's a small world after all" -- yeah, with a one-world government. Maybe if we play it backwards, it says that ... ?

Leslie

§

9-28-10

I suspect that from the very beginning, money from fund raisers like this is <u>never</u> intended to go to the touted worthy cause. The people who have bigger and better projects (like underground cities, perhaps?) know they can always rely on good old feminine dominance to create an <u>avalanche</u> of money into their hands, that then gets "tied up" in red tape somehow. Gee, isn't it strange. Just like the Federal Reserve was never intended to "stabilize the economy" but that's what people were told, this money was never intended to help Haiti, though that's what people were told. Everybody feels good while it's happening--the concerts, the donations, whatever--but the money never arrives. However, there's always tomorrow. I would love to be wrong.

Have to keep looking inside ourselves for sanity and love.

Leslie

§

The Invisible World – The Information Realm – Law of Attraction

The informational realm is like a place where all the blueprints are, the sacred geometry, the Flower of Life pattern for whatever is manifest.

3-20-08

For some reason, I feel sheepish about writing what I wrote. So, how about this for a vision of how things are:

Everyone is living in a world of their own creation and beliefs. There is never total consensus among everyone on even one single part of it. Alongside war, famine, and suffering, there is prosperity, thriving, and joy. It just depends on who you talk to. What if one's world consists of what one focuses on, and what if "In the beginning was the Word" expresses a universal principle at work in how things are created, then a big part of it all is choosing one's focus.

Leslie

§

3-22-08

At a seminar or lecture I attended once, the speaker told a story of a woman standing on a beach where thousands of starfish had been stranded on the sand and she was throwing starfish, one by one, back into the water. Someone came along and said to her "That won't do any good. It won't make a difference. There are too many of them." She continued throwing them back and said, "It makes a difference to that one. And that one. And that one." Your phrase "turning the screws of hopelessness" gave me a little shock. It made me think that it all turns on individual perception and interpretation.

"Turning the screws of hopelessness" in this case--pretending for awhile that there is a someone who wrote the letter addressing an audience--was relying on the tendency of people to imitate each other and to believe that their individual state depends on some general tendency in a group. But one by one, we can choose... something. And moment by moment, we can also choose, if we choose... to choose. Maybe I will take up saying kind of noble, inspiring things, like Eleanor Roosevelt, instead of just helping NWO do one part of their job--instead of advertising and promoting hopelessness!

Leslie

§

> *The mind is like a tuning-fork—it will set ringing around it whatever is in resonance with its own "vibration" or persistent thoughts and images.*

3-24-08

Dear All,

In my reading about how mind participates in creating experience, I came across an anecdote about Goethe. Goethe was familiar with these principles, which have lately been revived, popularized, and lumped under one principle with the label "Law of Attraction," which are described in books such as *The Master Key System*, and by writers and lecturers like Neville Goddard, Joseph Murphy, Charles Haanel, Wallace Wattles, Napoleon Hill, George Plummer, Emil Coue, and many others. The story is that when Goethe was working on some problem or other, he would spend hours in *imaginary conversation* with someone who was congratulating him on the successful outcome of his endeavor. In LOA circles, this is a well-known principle: create a scene IN YOUR MIND that encapsulates what you want to achieve, and then "LIVE" it, over and over again. Make believe. Pretend. Play. If all your ducks are in a row with the other principles involved, then just like apples falling to the ground, "reality" will orchestrate itself around what the mind is presenting as its focus. This is another way of saying "As within, so without" and "You can only see what you have in you" and "You will only see it when you believe it."

It has long been known that the mind is like a tuning fork --it will set ringing around it whatever is in resonance with its own "vibration" or persistent thoughts and images. Another way to say it is that the mind provides the blueprint for experience. It is also known that any strong emotion, including negative ones like fear and mistrust, will attract into one's experience the object of the emotion.

If Goethe knew about these principles and used them intentionally, I think it's safe to say that there were plenty of others like him that knew about them and that there are still people in high places that know about and use them.

The NWO is a pretty sophisticated bunch (maybe excluding Bush Jr) and from what I have read about mind control techniques, going even as far as traumatizing fetuses in the womb, I put nothing past them. I think they are smart enough to use everything at their disposal, including the science of mind, along with everything else --divide and conquer, distract and conquer, mis-educate and conquer, satiate and conquer, traumatize and conquer, etc.--to promote their aims. Maybe the main crop that they harvest from us is FEAR itself and submission to their control, so they keep marketing stimuli for fear, mistrust, and terror--in the form of various scenarios--so that we will be too scared and too much in "survival mode" to get in touch with the most majestic thing of all, the divine human spirit and consciousness.

By keeping the minds of a huge mass of people on the frequencies of fear, terror, deprivation, war, it gives them the added edge of law of attraction, with us contributing a "vibration" that promotes their purposes.

Leslie

§

5-15-08

Robert creates meaning in the world and all its details --they are talismans, reminders, and symbols of being present. That is what he chooses, that is what he believes, that is what he creates by choosing and believing it. And it's a lovely world.

LL

§

There is no shortage of good news. One simply has to talk about it, instead of the nasty stuff.

9-10-08

Dear all,

If we must talk in that terribly happy universe, we can talk about poetry and music and beauty and our water for energy devices, and our solar devices done with low-infrastructure technology, and about zero point energy devices like the Joe Cell and the contraption they have somewhere in Switzerland that some commune keeps almost totally secret, and about pyramid energy---there is a video on YouTube about a guy who experimented with harnessing pyramid energy; hung a length of wool yarn from the point of his 30-foot pyramid, and was thrown 20 feet or so when he touched it at the same time as the metal plate he wanted to connect it to. We could talk about using weather manipulation to bring water and bounty and balance to the whole world.

We could talk about the qualities of God, and about healing with intention, love & acceptance, and energy for all kinds of things that doctors and their hypnotized clients think are incurable. We could talk about healing and communicating with animals using energy psychology techniques. And the memory of crystals. Ever read about Marcel Vogel? Ever read about imprinting a crystal with the information field of an award-winning wine and then transferring that field to a mediocre wine, and experts could not tell the difference between the two wines? We could watch nature videos, and marvel at the infinite variety in nature.

Did you ever see the film about barrow birds, how the male bird collects things and very intentionally decorates his "bachelor pad" to impress lady birds who might wander by? Or one could study the Bible from the point of view of the "truth teachings" put forth by Joel Goldsmith, Florence Shinn, Neville Goddard, and others. Or one could learn the method of Stuart Lichtman for getting what you really want---who was probably influenced by the others, and combined their ideas with his deep background in engineering and business and artificial intelligence (Arintel), and his king of hearts and his intuition, and his wish to come up with a "model" for humans and how we work.

One could talk about the simplicity of the Budwig Diet for curing cancer (and arthritis and many other things) and the fact that the primary cause of cancer was discovered in 1931 by Otto Warburg---lack of oxygen to the cells (Nobel prize for it), which is correctable. One could read beautiful poetry, both old and modern. There could be writing workshops. Great for getting photographs and for learning to become a little more objective and unattached to the products of our gray matter. One could do intention experiments, like what was mentioned in *What the Bleep,* where 4000 people meditated and reduced the crime rate of Washington, DC by 25% or something like that. One could keep talking about wonderful things until we began to see God everywhere. Or one could keep quiet for long periods, until we started to see God everywhere.

There is no shortage of good news. One simply has to talk about it, instead of the nasty stuff. And as one talks about it, there will be more of it. As one talks about the nasty stuff, there will be more of that, too. It's a choice. There are people who claim to be eye-witnesses to the Indian rain dance, when rain would fall from a cloudless sky. The ancient so-called "primitive" cultures knew the role of mind in "creating reality." Read *The Mutant Message Down Under*, for example. They seem "primitive" to us probably because they could do so much with mind that they did not need all kinds of gadgets.

Plus, there are all kinds of other fun things to do! Like, sitting around a table, with various impressions on the table to stimulate the mind, and composing a group poem by going from person to person. Or, as they do in the Kiev center, sing together late into the night. And dance, there's always dancing.

Leslie

§

There is no shortage of good news. One simply has to talk about it, instead of the nasty stuff.

9-17-08

Well, while it all plays out, according to *The Handbook for Takeover,* I still think the solution is 1) wake up all young people, so they can't and won't be completely snowed, hoodwinked, and hypnotized into fighting "wars", 2) put the issuing of money back into the hands of a government that has the welfare of the people at heart, 3) also regarding money, set reasonable, lasting value upon it, and not let it be part of a system that can be manipulated with well-placed conspiratorial whispers in newspapers that send values crashing, 4) develop solar energy, zero-point energy, and water for fuel everywhere, 5) inform people about the fact that cancer can be easily cured with protocols such as the Budwig Diet and other things, 6) teach people energy psychology techniques with which they can clear and take care of many if not most of their psycho/emo/body issues, 7) do something about these central banks and monster conglomerates, and 8) teach everyone poetry, singing, and dancing from the old countries when they are young.

Leslie

§

10-1-08

OK, everybody put your play helmets on. We're going to play World (not House. World). Allen can be all the people in the Americas, Steven can be everything down under, John can be Europe, Charles can be Africa and all islands, and I'll be Asia and whatever's left.

So. Oh my, what is happening today? This morning, I woke up and went outside and I found a pile of money in a trunk that said "Eternally valuable because you are. With this money, bread will always be $1 and a gallon of gasoline will always be 30 cents, and beautiful houses will cost only $60,000. And each time someone is born, the government will issue them $100,000 to start life with, because they are worth that and much more!" I think I'll go tell my friends about it.

Americas, let's stop playing with the shrinking rubber money and issue our own, and let's make it so that nothing in the world can ever make it shrink, because no one else will ever convince us that it is worthless, because we decide for ourselves, because you and your goods and services will always have value, and the same for me, so our money will always have value, because it stands for us, and we will always appreciate (never depreciate) ourselves. Isn't this nice?

We will replace the dumb old monopoly money with ours, and suddenly everyone will have twice what they had before. Here is some stardust, and you know what happens when I sprinkle it on my countries? Suddenly, the rich people have a change of heart. They form banks, and if you open an account in one of those banks, it fills with their money for you to withdraw, perpetually, because their money is perpetually generated, and now they realize that it is more useful to share it. And they never become poor because they are experts at staying rich, and now everyone benefits from it. And look, here is a shovel, and when I dig with it, what do I find? Patents on free energy devices that are now public domain. Boy this is fun. Oh, you already have a car with one of those? How nice! And so does everyone else in your countries? That's wonderful.

Oh, look what else I found. This is a Puffball Informational Negativity Neutralizer. It causes certain machines and vehicles to malfunction before they can do anything at all, such as airplanes that are trying to take off to spread chemtrails. Have you seen any lately? Me neither. It's the PINN. Isn't it great? Oh, here's another thing: it's a powder that makes it so you earn lots of money doing something that you love.

> *One could keep talking about wonderful things until we began to see God everywhere. Or one could keep quiet for long periods, until we started to see God everywhere.*

Well, I'm going home now. I imprinted some water with the words "Forever Young." Don't you think I'm looking rather young and frisky lately? I'm getting taller, too. Tomorrow should be even more fun. I'll tell you about all the people that have cured amazing things. It's so funny, all these poor underinformed people scared into paying a lot of money to die. It's like that episode of Alfred Hitchcock, when two guys in a desert killed each other over a little water, because they didn't know the truck they had stolen was a water truck. It was too dusty to read. Oh well.

love, *Leslie*

§

10-13-08

It's amazing that "scientists" still publish such juvenile things about the impossibility of things that are a cinch for God. I guess they're just doing their part to keep people ignorant and controllable. Yet, in spite of all the social engineering of ignorance, which has reached fever pitch here in the end days of the Kali yuga, the cat has been let out of the bag all over the world, and understanding of "impossible" things is growing every day.

When Kiran developed the model for the software side of his Core Inergetix system, he came up with an understanding of what he called the "informational level," the dimension *before* energy. If his web site is still up, you can find his very good discussion of it there. Remember the phrase "In the beginning was the Word"? One could think of the "beginning" as the informational field, the place where all the blueprints are. Its signposts are "coincidence" and correlation. It's a very poetic "place."

Time and space are not coordinates, and one can obtain information from the field, as well as pass information to it, regardless of time, distance, or proximity. This is what is done by homeopathic medicine, radionics, and certain other modes. All systems of information that have to do with invisible correlation, such as the system of body types, all the coincidences that we call "shocks," astrology, feng shui, numerology, fortune telling, oracles, dowsing, law of attraction, science of mind, psychic perception, psychic healing, distance healing, and just plain old "coincidence," operate in this field. It interpenetrates all reality and all our lives at every moment.

It may be that, just as computers interface to the information-bearing world of the internet, DNA interfaces to the universal informational field, where the "plans" or maybe even the "divine plans" for everything and everyone are, along with information relating to many other things. Three anecdotes about the field:

1. A while back I read about Marcel Vogel, the fellow who developed liquid crystal technology for IBM. To make a long story short: One day when he was watching a saturated solution, he saw a flash of blue light just before the crystals formed. After a year of trying, he obtained a photograph of that flash of light, and what was in it? *A diagram of the crystals...*! In other words, straight from the informational field came the blueprint for the crystal at the point of incarnation, or incrystallization in this case.

I keep having this feeling that, even with this awful stuff happening, there is an equal or greater amount of good things happening and that, while many people who don't read what we read will be surprised at how bad it gets, we may be surprised at how good it gets.

Related to this:

2. In *The Mutant Message Down Under,* about the woman who went on a walkabout with Australian natives: One day a fellow broke his leg. Part of the tribal healing procedure was to run their hands over his leg, not touching, but very near. They said that *it helped the body reconnect to its perfect picture of itself*---the picture held in the informational field. One evening of healing, which included informational and vibrational modalities, and the fellow was walking the next day as if nothing had happened, according to the story. If this is true, it means that they understand the informational field and vibrational healing, as it appears many ancient cultures did, and live their lives in ways that reflect that understanding in many ways every day.

3. A 6-year old boy had such a traumatic relationship to eating that his growth was being stunted and he was starting to starve. His mother finally found a healer who knew EFT and other things. To make a long story short, the healer discovered that moments before the boy was born, the mother had said passionately to a doctor "I have become so fat don't ever let me eat again! Don't ever let me touch food again!" Bingo. So, the healer accessed the field, explained to the boy (who was not in the office) what had happened, then she did distance, surrogate EFT for the grandfather, the mother, and then the boy, all around eating issues, especially the birth event. The mother went home, and one hour after she arrived, *for the first time in his life,* the boy said to her, "Mom, I'm hungry. Would you give me something to eat?" He grew completely normal and healthy, from that point on.

Regarding the informational field and how we experience our lives: At every moment, and at the same time, we are broadcasting towers sending out a signal, and we are also receiving antennae for whatever resonates with our vibration---it's like we are piano strings that set vibrating anything or anyone tuned to the same note. Bruce Lipton talked about this in that long lecture on DVD. We see only the things that resonate with our note, only the things that can pass through our filter. But, in addition to that, something "out there" reads the vibration instantly, and immediately orchestrates objects, events, conditions, etc. for us to experience immediately, or eventually, as the corollary to our dominant vibration. Sometimes it's wonderful, sometimes ironic, sometimes tragic, sometimes funny, sometimes commonplace, but if one does not know that this principle of correlation operates at all times, and one is not paying attention to one's most persistent thought/feeling output signal, the correlations pass "invisibly" without our noticing, and we tend to blame everything but ourselves if it's not fun.

The people (or whatever) who have orchestrated the "war on terror" and the so-called "inevitable collapse" of the financial system know this, most likely. So they instill fear in everyone everywhere. Then, this massive vibration of fear puts a blueprint of calamity into the informational field, and helps to attract the things that are feared, making it easier for the perpetrators. So... it's important as individuals to not worry, and be happy, like all the children and animals that don't listen to the news... because mind creates blueprints.

love, *Leslie*

§

Reality is, in reality, a very magical thing, an ongoing magical thing, and we participate in it, but mostly in an unintentional way.

12-3-08

"Show me what I am, *truly.*"
That is the command implicit in every effort to be present, in every work 'I,' in every *BE*.

"To see a world in a grain of sand,
And a heaven in a wild flower,
Hold infinity in the palm of your hand,
And eternity in an hour." --- William Blake

I don't know how to put this elegantly, so I will just put a few chunks of ideas together here:

Everything, everyone, every happening, all the space in between, and anything left out of the above, is made of only one thing---the ultimate oneness of the One Somethingness. Because of that, everything, everyone, every happening, and anything not mentioned, is a hieroglyph, or hologram, of that One Something-or-Other, somehow secretly invested with the same divinity. Therefore, to make as many things as possible a reminder to be PRESENT is about as direct a connection as we can devise, for now, for *deciphering* the hieroglyphs of that fabulous Whatever-It-Is.

because we can only decipher the signs
when we ***are***
what we ***truly*** *are---*
God,
Love,
Intelligent light,
Presence.

... according to what I and we have read from various well-respected people from various spiritual traditions, and according to what I have experienced to a tiny degree. But you know what else? I would take Calleman with a grain of salt. He could easily be a pawn in the great game of getting everyone to *BELIEVE* that disaster is coming in 2012. And "Let it be done unto you according to your belief" is not just Christian chicken feed. It is a big, whopping, universal principle that the power brokers of the media use to their own great advantage.

In spite of everything we know, I still have the feeling that something incredibly wonderful is in the works.

Leslie

§

1-21-09

They talk about a reward area of the brain, and a punishment area, but he does not speak about how programming may influence what one perceives as deserving of reward or punishment. Gurdjieff, for example, was told by his grandmother to never do what others do... so I wonder what part of his brain lit up whenever he disagreed with the group.

L

(*The foregoing was in response to the following email* -- *Ed)*

From: ....
Subject: Why so many minds think alike - or how Bush became president
Date: Wednesday, January 21, 2009, 5:46 AM

Why so many minds think alike?

§

4-1-09

Whether we are conscious or not, individual and collective thought/vibrations DO guide how it all works, or rather, how it all manifests. That is the "secret" in the movie *The Secret.* To have a mind--*whether one is conscious or not* --means that we are renting a piece of God's mind. Renting a piece of God's mind, with all its creative power, but at a junior level. "In the beginning was the Word" (in God's mind)... so we should learn to choose our "words" well, in every way.

It is not a contradiction to the idea that "man cannot do" because it's true: alone, man cannot do anything, but God (the absolute, the oneness, whatever) can and does do everything, and I don't think we are as cut off from certain aspects of God as Ouspensky believed. What we CAN do, though, is choose our thoughts, our words, our focus, and it applies to everything in life as well as to choosing work 'I's to guide us to presence.

The "fragments of an unknown teaching" that we got through the Fourth Way focus on creating consciousness, but there is more that demonstrates that reality is, in reality, a very magical thing, an ongoing magical thing, and that we participate in it, but mostly in an unintentional way.

§

6-7-09

Dear friends,

Here is a link to a talk and interview with Bruce Lipton (the fellow who wrote *The Biology of Belief.* Go to it soon, because the page may be taken down.

(Website link, originally provided here, has been removed. -- Ed)

In this talk he elaborates more on how our beliefs and perceptions affect the expression of genes, our biology, and our lives, ties in to 2012, and describes the roles of "surface" or "conscious" mind (in ordinary parlance) and the subconscious mind. His statement about that is:

The role of the mind is to create congruence between our beliefs and our lives/experience.

(I think we know of a third part---higher mind experienced through higher centers, but he does not talk about that.)

He also talks about the idea that humanity is undergoing evolution, or a possible evolution, and uses the metaphor of the caterpillar and butterfly to describe what is happening, when inside the chrysalis the old caterpillar turns into soup, while "imaginal" cells then play the role of reorganizing that soup into a new and better being.

Very interesting, very upbeat, very optimistic. Enjoy.

Leslie

§

Sexual energy is lawless and it ventures into all the "forbidden" worlds, whether they are higher or lower.

10-4-09

It seems that scientists, because of their huge investment in advanced education from the powers that be, are limited to rather pedestrian and halting approaches to understanding mind and consciousness, but I guess little steps are better than none. When it comes to describing the capacities of the mind, I find that such authors, teachers (with a small 't'), and professionals as Florence Scover Schinn, Neville Goddard, Joseph Murphy, Emile Coue, Jose Silva, Esther & Jerry Hicks, Stuart Lichtman, Jill Taylor, and many others, are much better sources for discovering what are the capacities and good uses of mind, intention, and decision, and the impact of contradiction and negative and positive emotions on our day-to-day experience.

I keep having this feeling that, even with this awful stuff happening, there is an equal or greater amount of good things happening and that, while many people who don't read what we read will be surprised at how bad it gets, we may be surprised at how good it gets.

Trying to read through scientific studies gives me the feeling that I am wasting time and mental energy on the findings of a clumsy approach to something surpassingly elegant and mysterious: the mind and its operation in both the visible world and the invisible world of correlation and attraction.

The visible/physical world, even electricity and other measured energies and radiations, are a small fraction of reality. As we know, Dr. Bruce Lipton is one who leads the way out of the Newtonian point of view. He was one of the lucky ones who was led into a broader understanding when at some point, he was able to SEE what was right in front of him in his Petri dishes, and then put it together with flashes of insight and the many other clues he had from researchers whose work had been ignored and downplayed. Even Einstein, the only scientist whose name is a household word (because he contributed to our most exalted technology: the big bomb) acknowledged the importance of intuition. Scientists trying to stay professionally alive are shackled and hindered by scientific dogma and clumsy research tools like "scientific method." But even that method, in spite of its limitations, sometimes provides surprisingly provocative evidence of something more.

hugs to all, *Leslie*

§

3-2-10

With the principle of "as above, so below" operating, then a major bobbing up and down of land masses, with totally redrawn coastlines, etc., etc., could be accompanied by similar events in the minds and hearts of both god and earthlings. Land masses rising ... unconscious beliefs/attitudes coming to light? New capacities coming to light? Previously hidden or little used parts of our psyches coming to light and creating completely new and different ways to live? Parts of land masses sinking under water ... the heavy, destructive aspects of our psyches dissolving and sinking to hold up the ocean for awhile?

I agree that these things are not done by earthquakes alone, but rather by the passing of planet-sized comets that have the mass to yank continents out of place. So if NASA has reported such a shift in the axis ... it could be laying the groundwork for something else that will be falsely or half-falsely attributed to the combined disturbances of this recent earthquake and __________?

Ancient paradigms of healing and manifesting life are being excavated, dusted off, and innovated upon to make them accessible and friendly to everyone. The future, as previously predicted, could be changing due to so much TRUTH information circulating globally, and due to so many people on the planet trying to be present, and trying to heal their beliefs and psyches and world views in whatever ways they understand. Although they and we represent a tiny fraction of the total population, the numbers may be reaching the tiny but effective ratio by which, according to some people, the positive forces counterbalance the negative ones.

LAL

§

4-2-10

THIS IS GREAT!

LET'S START WRITING OUR OWN NEWS,

every day!

Wonderful news, news that makes our hearts leap with joy! News that brings sighs of relief, fresh breezes of simple truths, simple solutions, simple presence. News that opens doors and brings in wonderfully fresh new possibilities.

Because, the First Step in Manifestation is: Twinkle in the Eye. This really puts a twinkle in my eye. How about you?

You see, the mind should not be used only like a donkey on a rope plodding along after everything that "happens," which we feel so morally bound to describe as "the facts, ma'am, nothing but the facts." THINGS ARE NOT WHAT THEY SEEM. Facts may not even BE facts! They may be <u>manifestations of mind, or collective mind</u>. So, better to use our MINDs the way Original Mind uses it---to create beautiful, joyous hints and reflections of the One, somehow, like with presence, good news, and joyous tidings.

hugs, *Leslie*

(The foregoing was in response to the following "April Fools" joke. -- Ed)

Sudden Mass American Enlightenment Puzzles Congress and Causes National Security Alert

Thursday, 01 April 2010

President Obama, U.S. congressmen and their aides held an emergency four hour-long meeting Wednesday to decide what action to take to stem a sudden and unprecedented mass enlightenment of the US population that threatens to destabilize the Government and create a fiscal crisis.

Millions of citizens jubilantly celebrated their new awareness as they suddenly realized they have been lied to all their lives on just about everything by their psychopathic, conscienceless leaders of major corporations, government and security agencies.

Residents in all major cities across the United States awoke Wednesday morning complaining that they were experiencing an inexplicable mental clarity that was causing them to realize the way they were about to spend their day was utterly pointless, a potential health threat and would only add energy and finances to a powerful wealthy elite that had been ruling their lives since as long as they could remember.

Congress was thrown into chaos when a majority of staff decided to simply stop showing up for work until their bosses resign; in a move of solidarity, energy companies turned off the power to government buildings and military bases, while water companies shut down their water supply; repair people refused to carry out repairs for the politically well-connected and television network staff refused to report the lies they are routinely ordered to tell by their editors.

§

5-10-10

Thanks very much, Marion. The sweetness of that dog reminds me of yours.

And Charles, here are some thoughts:

What if it means that to awaken, one must be a dog or cat or elephant in one's ninth lifetime? Maybe the original line in the Bible was, "To enter the kingdom of heaven, you must be as a little puppy"--or kitten or dolphin-- but the authors just couldn't bring themselves to write it, because attitudes towards animals were even worse then than now.

What if the "intelligence ratings" are exactly the opposite of what we think, that the animals--not we--have superior intelligence and emotional being? Look at everything they put up with--our cruelty, our ignorance of their needs, the prisons and boredom we create for them in their domesticated lives. What if, in those moments when they seem "so intelligent" or "so emotional," we are not looking down, but UP at them, and catching a partial glimpse of something that is beyond and above us, but we are so blinded by all the propaganda about Our Intelligence that we don't see it?

What if not being able to perceive the love and intelligence of animals is a form of racism, or species-ism? When we do not recognize the humanity of a different race of human, we treat them like . . . animals . . . because we don't recognize the "humanity" of animals, either, and so, of course, see nothing wrong with treating them like . . . animals.

We have *the potential* to be more intelligent, and therefore deserving of the admiration that we imagine the “more intelligent” animals have for us---what with having not just three, but four, centers, and pineal glands, and opposing thumbs, brains that can use symbols (which animals can do, by the way, if they are given a rich educational context, like the ones we enjoy, in which we *learn* to . . . use symbols), and tongues that can speak languages, etc., etc., etc., but our perceiving apparatuses really are limited, and there are many languages that require not intellect but heart, which animals have a lot of (especially giraffes--25 pound hearts!), and intuition, and the being-ness of being connected to the great, the mysterious, the totally within and without, the fabric of all, the one and only, the simply profound and profoundly simple and ever present--One.

Leslie L.

§

The powers-that-be know that people are satisfied with having emotions about it all, rather than knowing or understanding what's really going on.

Miscellaneous Subjects

> *Although they and we represent a tiny fraction of the total population, the numbers may be reaching the tiny but effective ratio by which, according to some people, the positive forces counterbalance the negative ones.*

8-2-08

I like the metaphor, the "Exit Topic Button."

In something that I read about mind control/social engineering long ago on the www.educate-yourself.org site, it talked about the "slide," which is a response to certain topics that the powers-that-be don't want people to think about or investigate. People are indoctrinated to respond to these forbidden topics with a "slide"--and maybe other things, but this is the one I remember-- which is basically their "Exit Topic Button."

So for example, this social engineering has been applied to the word "conspiracy." It is a word that evokes that response in people, and I think we have all seen it. As soon as one starts suggesting that something is going on behind the scenes, the slide response is triggered: people get a certain kind of smile, a certain look, they may get a little agitated, and then they say something that nearly always includes the phrase "conspiracy theory," which makes them feel safe and intelligent and worldly-wise, and also tickles them with the opinion that if YOU are thinking about "conspiracy theory," you must be crazy, paranoid, and delusional.

And that's the end of the topic for them. Maybe a few "exit remarks," but no thinking, no letting in any information, because of course, THEY don't want to be crazy, paranoid, and delusional. The movie "Conspiracy Theory" was pretty good for showing a little bit about certain mind control tactics, but of course there was the half-truth pill that viewers swallowed along with it, which was that the government is essentially made up of good guys that want to get the "bad-guy rogues" that use mind control on people.

In the FOF, I think we have gotten a good dose of indoctrination about various things, along with the gems of esoteric knowledge and the gems of the experiences of higher states. It's interesting to try to sort them out. I'm not saying it's an awful thing. We were all willing to absorb something very radical and new, and we believed that it had to be swallowed whole. But now, as always, it takes true presence--true experience of higher centers and not just words and talk--compassion, forgiveness, and humor about oneself and other people to see things in right perspective and keep an even keel.

Leslie

§

9-9-08

Many weeks ago, a student who thought I was good at wordplay came and asked me what "shocks" I could interpret from the name "Barack Obama." Between the two of us, we came up with "barracks----oh, bummer!" Barracks as in military rule after another phony 9/11, and "oh, bummer" as in, like, "wow, what a drag."

Leslie

§

9-10-08

Dark matter and dark energy may perhaps be certain regions of the Mind of God, and perhaps they seem "dark" only because humans operate in such a narrow bandwidth of perception. If we could tune in, it might be blindingly and celestially bright. And while this may sound like drivel to Newtonians, the concept that everything comes from and is made of God, which is now part of my model of the universe, is what I deduce from much of the reading I've done in the last two or three years, and from my experience with the energy psychology technique, EFT (emotional freedom techniques), and from other experiments.

Leslie

§

Next to sunlight, water, and empty space, sex energy is the most abundant thing on the planet.

11-30-08

Maybe there was a migration of Greys and others from Mars to Earth, just before the comet-that-became-Venus fly-by that stripped Mars of most of its atmosphere and water. Would that jibe with Beelzebub's tale?

Leslie

(The foregoing was in response to the following email -- Ed)

From: ....
Subject: NASA Mars photo leaked - wood found on mars!
Date: Friday, November 28, 2008, 6:57 PM

NASA Mars photo leaked - wood found on mars!

§

6-7-09

Sexual energy is lawless and it ventures into all the "forbidden" worlds, whether they are higher or lower. Having sexual drive in the instinctive center is no guarantee that the third higher center---the sex center---will develop, or has developed. Balanced development of sex center seems almost as elusive as the other higher centers

Leslie

§

1-25-10

Dear everybody,

Thanks, Charles, for including me in the mailing list. John and Allen might have thought that I don't want to hear anything about the sequence or keys, or evidence of schools based on finding the keys, but that's not the case anymore. I am proofreader, formatter, and sometime editor of Thoughts from the Teacher, and it has been a gift in that first of all, it gave me a crash course in the keys and the sequence at a crucial moment, and secondly, by sheer repetition from reading and sometimes writing the Thoughts issues, and by serendipitous conversations with various students, I no longer scoff at or argue with the sequence. On the contrary, I understand it, and use it, sometimes "by the book" and sometimes with variations, as a simple, direct way to try to be present, or to witness presence, or to make a clearing for presence, or to accumulate energy for presence, ten breaths at a time. One student calls it "an upgrade to Vipassana," which is also a form of breath meditation for clearing and creating presence.

That's interesting that Bacon's group was called the Spear-Shakers, and that Robert has keyed the Pharaoh's spear as referring to the 30 work 'I's.

hugs, *Leslie*

§

7-16-10

(Referring to "Crop Circles" -- Ed.)

If the designs are "beamed" onto the ground, perhaps they are beamed all at once, the way images are projected onto a screen by one light source.

And if they are intended as messages, they apparently have little to do with words. Maybe they are intended to insert a certain particular vibration or resonance into the "field"--the crop field as a symbol of the vibrational field--of humanity. Things like that can be powerful and healing, or regenerative--whatever the intention is. See Ibrahim Karim on Biosignatures for discussion of this phenomenon. Maybe they are part of the good forces that, according to some people, are working or fighting on our behalf, while dark forces are laboring to bring about our demise. And maybe their effects, if they are "good," are not felt as something new, but simply as the fact that we are not yet all dead, and more and more information is coming out about the many ways in which we have been deceived.

Leslie

"Our original Buddha-Nature is void, omnipresent, silent, pure; it is glorious and mysterious peaceful joy—and that is all." [Huang-Po]

§

Health & Healing

It seems that the most powerful parts of the immune system at all levels are love and the mind and the intention to be healthy and whole.

4-8-07

Dear all,

If Kiran's site is still up (probably is) he has a lot to say about this. His material is very interesting because he makes a distinction between "energy" as it is talked and talked about and another dimension that precedes energy, which he calls "the information realm" or "CoRe space" (also called the realm of synchronicity).

The informational realm is like a place where all the blueprints are, the sacred geometry, the Flower of Life pattern for whatever is manifest.

You can get information from the blueprint, and you can pass information to it. That is what CoRe Inergetix does, that is what psychic healers do, that is what oracles like the I Ching do, that is what we do without knowing it throughout our lives.

It is ordered by principles different from the realms of energy and matter as we readily perceive them. Firstly, in that realm, everything--absolutely everything organic or not organic or otherwise-- is "connected." In other words, "As above, so below. As within, so without. As the universe, so the soul." Secondly, thirdly, and fourthly, time and space are not parameters there (hence, healing and other operations at a distance), symbol is very powerful, and things occur "coincidentally" when they have a similarity of some kind or a "resonance" with each other because of a similarity of some kind in the perception or experience of someone.

Stephen sent a link a while back about research on DNA that says that DNA is not hard -coded but rather an antenna for information... it is a changeable blueprint that we can see under a microscope. The research also said that it is structured the same as, and can be changed by, language. They said that it requires certain specific frequencies to accompany this; however, my experience with EFT (emotional freedom techniques) leads me to think that language and intention coupled with stimulating the energy meridian system can also communicate with it and quite possibly change it to some degree.

There are many interfaces to the blueprint realm. Homeopathy is one. White noise of a computer data line is another. Others are "radionic" programs and devices like Kiran's. Others are through divination and dowsing devices, like pendulums, oracles, "water witching" and so on. The human body itself is a tool of divination if one pays attention to it. There are various energy therapy and change techniques that seem to provide interface to the blueprint and the energetic realms at the same time, judging by their miraculous results in thousands of cases. EFT, which can be practiced and can produce effects from any "distance" (distance does not exist in the informational realm) is one of those.

From what I know so far of different energy therapies like Hooponopono Healing Prayer, EFT and TAT, all of which are gentle, powerful, and versatile healing/change techniques, the essential ingredients are:

- Intention and attention
- Focus on whatever the problem or desired change is (in words or just with your attention)
- Love, acceptance, forgiveness of self and others
- Interface to the blueprint (directly, mentally, or symbolically touching or tapping points in the energy meridian system or having a picture or description of the target)

Hooponopono is based on the Hawaiian teaching that you create everything in your world. Sound new age? They have known it for centuries. There is a story of an MD who agreed to be the resident doctor at a mental ward for the criminally and dangerously insane. No other doctor had lasted longer than a month. He never spent time with any of the inmates, although he had their files and photos. Within four months or so of his arrival, everyone in the ward was cured. Even the staff started to like being there. Eventually, the inmates were all released and the ward was closed.

Someone interviewed him later on and asked him what the heck did he do. He explained that whatever is in your world is in you and part of you, and you must heal yourself. "So what did you do?" the interviewer asks. I said "I'm sorry, please forgive me, I love you" over and over.

It seems that the most powerful parts of the immune system at all levels are love and the mind and the *intention to be healthy and whole.* That is the elephant in the living room that mainstream indoctrination about health, sickness, drugs, and surgery does not want us to see or use.

love, *Leslie*

§

Everyone is living in a world of their own creation and beliefs.

9-3-07

Hello John,

So, you are experimenting with homeopathic coffee and croissants. You are taking your first intentional steps into the realm before energy--the informational dimension, the realm of synchronicity, the blueprint dimension.

We are always having intercourse with this realm, but we have not been trained to recognize it, nor do we receive training in its principles and axioms, unless one works in certain disciplines. The axiom "As above, so below; as within, so without; as the universe, so the Soul" refers to principles that order the informational realm and how it affects the other things that we are more familiar with--energy, space, matter.

One can receive information from this realm, as psychics and all other oracles do (I Ching, Tarot, tea leaves, all our "shocks" and coincidences), and one can also *send* information to this realm to make changes, healings, corrections, that then appear in the lower dimensions, i.e., the body and circumstances.

One example you may already know is homeopathic medicine. The little pills and drops have NO substance or "electromagnetic energy" in them. All they have is the *imprint of information of the remedy*, and when you deliver the right information to the body, if that's what it needs, the response is "miraculous", which, as we know, means the operation of the laws of a higher realm in a lower one.

Words, images, and symbols are powerful because they act in the informational realm, where the blueprints, the Original Ideas, the matrices of being, are, and because they often evoke emotion at the same time, which is like a transport vehicle for the message. If getting the wrong X-ray and being told you have cancer can give you cancer, the opposite is also true. People have effected astonishing healings for themselves by persistently believing and stating to themselves that they are whole and healthy and well. People also accomplish remarkable things because they persistently believe they can and act in support of their belief. They also fail to accomplish other things because they persistently believe they can't.

Energy healing modalities, like EFT (Emotional Freedom Techniques) and TAT (Tapas Accupressure Technique) are as effective and versatile as they are because they work with mind, intention, acceptance, and energy. They work in the informational AND the energetic realms.

Leslie

§

3-30-09

In circles where people are versed in systems that work in the invisible realms of information space and energy (feng shui, Chinese medicine, all divination arts, "distance" healing, radionics, etc.), it is expressed as "the pattern of the whole is contained in every one of its parts." As above, so below. As within, so without. As the universe, so the soul. Hence, divination: from reading the palm of a hand, to listening to the cries of birds, to Tarot cards, to the I Ching, to shocks from C Influence, to astrology, to body types, to etc.

Also, it is becoming more and more clear to me that what we have been taught to call "reality" is actually only the *readout* of information patterns and energy patterns, just like a computer monitor gives you a readout on what you do with the energy in the computer. So a method, exercise, system of ideas, or discipline that allows you to change those patterns in your body and your life--and there are many very accessible tools now in the areas of healing and self-development---can produce profound results in the readout---our "reality"---with a delay that depends on partly on lawful relationships between those things, and partly on how much resistance to the new pattern remains.

(The foregoing was in response to the following email -- Ed)

From: ....
Subject: Proof of Holographic universe?
Date: Monday, March 30, 2009, 10:28 AM

Folks,

Here's a pretty interesting article, from New Scientist Magazine.

Our World May Be A Giant Hologram

§

10-10-10

Dear friends,

That is Charles putting words into my mouth. I don't think I have ever said to anyone "let's keep the conversation on a high level" because I'm usually the last one to be able to do that! Usually it's the other way around---someone censoring (or photographing) me.

But it's true, I need to concentrate on healing from surgery right now, and should take a break from all the salacious news about the wicked shenanigans of our brilliantly wicked overlords who harvest and wish us harm daily. But we smile and get by anyway, don't we?

love to everyone,
Leslie

§

(The following is Leslie's final contribution to the discussion group. She passed away on May 13, 2011. – Ed)

10-11-10

Dear friends,

My last email was too blunt. I thought it would be taken as "conversational banter" which is what Charles and I often have, but with email people often "hear" a different emotional tone than what we think we wrote in, so I just want to say that by "putting words into my mouth" I know Charles meant the very best for me. He and others have been incredibly, incredibly helpful to me during the last few weeks, and more, So keep up the research, everyone, and keep an eye on the world for me while I hibernate in my healing cave.

love, *Leslie*

§

"If I Can Dream"

There must be lights burning brighter somewhere
Got to be birds flying higher in a sky more blue
If I can dream of a better land
Where all my brothers walk hand in hand
Tell me why, oh why, oh why can't my dream come true

There must be peace and understanding sometime
Strong winds of promise that will blow away
All the doubt and fear
If I can dream of a warmer sun
Where hope keeps shining on everyone
Tell me why, oh why, oh why won't that sun appear

We're lost in a cloud
With too much rain
We're trapped in a world
That's troubled with pain
But as long as a man
Has the strength to dream
He can redeem his soul and fly

Deep in my heart there's a trembling question
Still I am sure that the answer's gonna come somehow
Out there in the dark, there's a beckoning candle
And while I can think, while I can talk
While I can stand, while I can walk
While I can dream, please let my dream
Come true, right now
Let it come true right now
Oh yeah

Elvis Presley

http://www.livingpresence.com
http://www.bluelogic.us

Made in the USA
Monee, IL
08 November 2026

67377479R00042